D1070823

STRUCTURE

EXPLORING EARTH'S INTERIOR

STRUCTURE

EXPLORING EARTH'S INTERIOR

ROY A. GALLANT

BENCHMARK BOOKS

MARSHALL CAVENDISH
NEW YORK

For Martha

Series Consultants:

LIFE SCIENCES AND ECOLOGY
Dr. Edward J. Kormondy
Chancellor and (professor emeritus) of Biology
University of Hawaii—Hilo/West Oahu

PHYSICAL SCIENCES
Christopher J. Schuberth
Professor of Geology and Science Education
Armstrong Atlantic State University
Savannah, Georgia

Benchmark Books
Marshall Cavendish
99 White Plains Road
Tarrytown, NY 10591-9001

Library of Congress Cataloging-in-Publication Data
Gallant, Roy A.
Structure: Exploring Earth's Interior
    p. cm. — (EarthWorks series)
Includes bibliographical references and index.
Summary: Describes the formation of Earth, the composition of its surface and interior, and the effects
of earthquakes and volcanoes.
ISBN 0-7614-1368-5
1. Earth—Internal structure—Juvenile literature. [1. Earth—Internal structure. 2. Geology.]
I. Title. II. Serries
QE509 .G35 2002
551.1—dc21   2001043858

Photo research by Linda Sykes Picture Research, Hilton Head, SC

Cover: E.R. Degginger/Photo Researchers
The photographs in this book are used by permission and through the courtesy of: John Burnley/Photo
Researchers: 2–3; Erwin and Peggy Bauer/Bruce Coleman: 6; Space Telescope Science Institute/NASA/
Science Photo Library/Photo Researchers: 10; NASA/Science Photo Library/Photo Researchers: 12;
ChromoSohm/Sohm Photo Researchers: 16–17, 32–33; Mark Burnett/Photo Researchers: 18–19; Ken
Johns/Photo Researchers: 23; Dr. Karen Von Damm, University of New Hampshire: 24, rear cover; David
Parker/Photo Researchers page 36; David Hardy/ Photo Researchers: 42; Roger Ressmeyer/Corbis: 46
(top); David Hardy/ Photo researchers: 46 (bottom); Bernhard Edmaier/Science Photo Library/Photo
Researchers: 48; Jim Corwin/Photo Researchers: 50–51; John Meehan/Photo Researchers: 52–53; Dietrich
Rose/OKAPIA/ Photo Researchers: 56–57; Charlie Ott/Photo Researchers: 60–61, rear cover; Francois
Gohier/Photo Researchers: 62–63; Jim Steinberg/Photo Researchers: 64–65, rear cover; Bernhard
Edmaier/Science Photo Library/Photo Researchers: 67, rear cover.

Series design by Edward Miller.

Printed in Hong Kong

6 5 4 3 2 1

CONTENTS

INTRODUCTION

Earth has been around producing changes much longer than humans have been here trying to understand why. Some ancient Greek thinkers dismissed volcanoes simply as "burning mountains." The great Aristotle said that earthquakes and volcanoes were caused by winds trapped in underground caverns that built up until they finally burst out.

It wasn't until people became curious enough to actually observe and describe the shaking of the ground or the rumbling of a volcano that a scientific attempt to understand what was happening began. One such person was the Italian painter and inventor Leonardo da Vinci, who lived in the late 1400s. On finding fossil shells of marine animals high in the mountains of northern Italy, he could have merely shrugged and moved on, instead of rethinking the prevailing ideas about fossils. One was that they were crumbs spilled by travelers passing through the area. Another was that they were the work of God and so need not be questioned or explained. But explain them he did by correctly saying that the mountain must at one time have been part of an old seafloor that somehow was thrust up high above the ocean.

FROM WINDY CAVES TO SATELLITES

It has been in the spirit of such inquiry and reasoned thought that scientists have come to understand at least some of the workings of Earth as a planet—how it is put together and what it is made of. Today thousands of scientists the world over are drilling into the seafloor, measuring ground tremors from earthquake stations, drilling into glacial ice, making observations from artificial satellites, and planting sensors around the base of volcanoes to find out what makes Earth tick. It is remarkable to think that a

"Fire rock" is one name for the lava that spills out of volcanoes and coats the land with glowing tongues of molten rock, as during this eruption of Hawaii's Kilauea volcano. When the lava cools quickly, it becomes volcanic glass known as obsidian. Solidified lavas are an example of igneous rock.

research satellite traveling 4 miles (7 kilometers) per second 500 miles (800 kilometers) above Earth can measure the displacement of the ground at the two edges of a fault, or crack, to an accuracy of only a fraction of an inch.

Only in the past several decades have these investigations begun to tell us something about the *structure* and *composition* of planet Earth. We have learned that Earth has three major regions: a thin *crust* of lightweight rock beneath which is a much thicker region of heavier rock called the *mantle*, and beneath that a large *core* of mostly iron and nickel. Since it's impossible to travel to Earth's center, scientists have had to use indirect means of probing the planet's interior. By setting off small explosions, for example, they have been able to measure the thickness of *sediments* on the seafloor and learn about the rock that lies beneath.

Geologists have even been able to discover what triggers earthquakes and what makes volcanic mountains sometimes blow themselves to bits. But the ultimate causes of these events are hidden, out of sight and out of reach, deep beneath the crust. Some of the forces that build mountains are known, but the root causes, like the roots of the mountains themselves, are also hidden far underground. Nevertheless, what geologists have discovered about the workings of planet Earth, over the past fifty years alone, would astound and bewilder Aristotle, as it has astonished the scientists who are making those discoveries today.

ONE

FROM DUST TO EARTH

If we could dig a hole down into our planet's center, it would be about 4,000 miles (6,400 kilometers) deep. We could then study all the material we removed and know exactly what Earth is made of and how it is put together. Unfortunately we aren't able to do that. The deepest hole anyone has ever dug is in Russia and goes down 7.6 miles (12.2 kilometers). This well is being dug for scientific research.

The red-hot rivers of *lava* that spill and gush out of volcanoes tell us a lot about the rock material buried several hundred miles inside Earth—how hot it is down there, the chemical makeup of the material, and a wealth of other information as well. As the lava comes pouring out to meet us, it's easier to sample and study. When still underground, lava is known as *magma*. We have used other methods to find out about all the materials that lie beyond the magma and learned a lot about Earth's interior in the process. It turns out that in addition to the great underground pools of magma, our planet has rock layers heavier than the densest mountain and an enormous ball of iron forming its core. But where did all these materials come from in the first place?

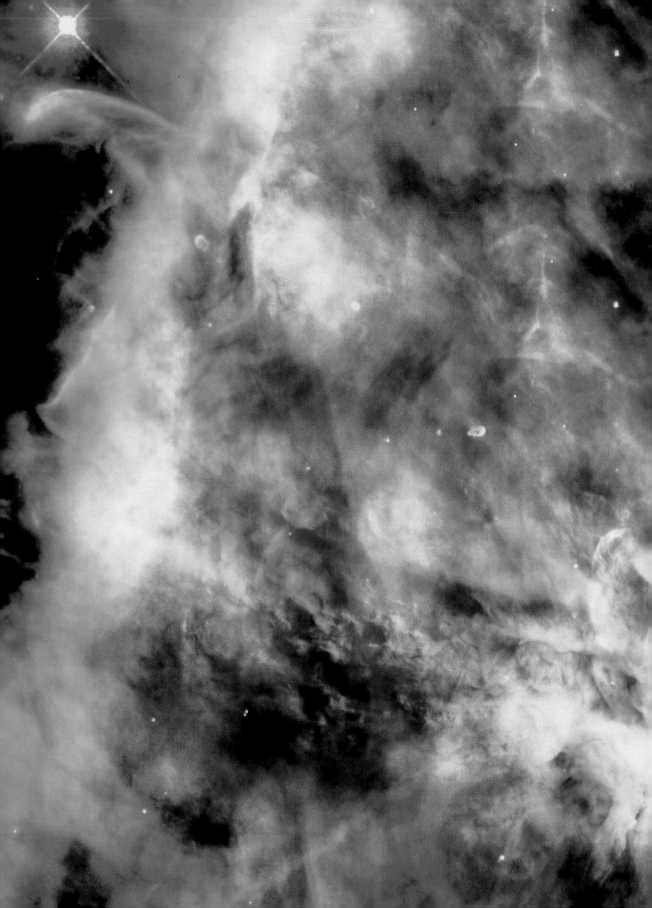

LUCK FROM A DYING STAR

To answer that question we have to turn the cosmic clock back five billion years. We also have to travel to a place near the edge of our galaxy, the Milky Way, where a giant cloud of hydrogen and helium gas hovers. Stars are known to form out of clouds such as these, and the Sun is one of billions of stars to do just that. For millions of years the Sun-cloud had been collecting matter as its gravity pulled in large amounts of the gas and dust floating through space. Eventually, the cloud began collapsing in on itself and heating up as its matter became packed ever tighter in the central region of the cloud.

That cloud that was to become the Sun shared its cosmic neighborhood with other stars, some the same age as the newly forming Sun and others much older. Most stars have cores that are so hot and under such great pressure that the hydrogen in them is crushed and packed so tightly it forms helium. Other chemical elements such as oxygen and carbon atoms are made as well. However, especially massive stars are element factories that produce chemical elements heavier than carbon. These include iron, nickel, gold, and uranium. Each of these heavy elements is created in only seconds when the massive star ends its life by exploding violently. The atoms it has made are then cast off, some in the form of elements clumped as molecules called cosmic dust. They then travel through space.

The cloud of heavy elements sent off by one such *supernova*, or dying giant star, just happened to cross the path of, and mix with, the hydrogen-and-helium cloud out of which the young Sun was forming. Enriched with those heavy elements, the Sun-cloud continued to collapse until eventually it heated into that raging inferno we see as the Sun today. Meanwhile it had begun to spin and expel matter that formed a huge disk of material revolving

The great cloud of dust and gas called the Orion Nebula is one of thousands of nebulae in which star formation is taking place. Stars like our Sun form out of these great clouds, shine for a few billion years, and then end their lives as dark, dead embers.

about the Sun. The disk material included gas, clumps of atoms in the form of dust grains, ices, and rocks of silicate materials mixed with iron and other metals. For a hundred million years these clumps of matter—ranging from the size of golf balls to houses to entire mountains—collided and stuck together as objects called *planetesimals*.

As some planetesimals collided, they shattered. Others were nudged into new orbits that sent them to a fiery death in the Sun. Still others assumed courses that sent them away from the disk as cosmic exiles. The gravity of the more massive clumps enabled them to sweep up less massive ones and so grow larger. If they were larger than about 200 miles (320 kilometers) across, gravity pulled them into the shape of a sphere.

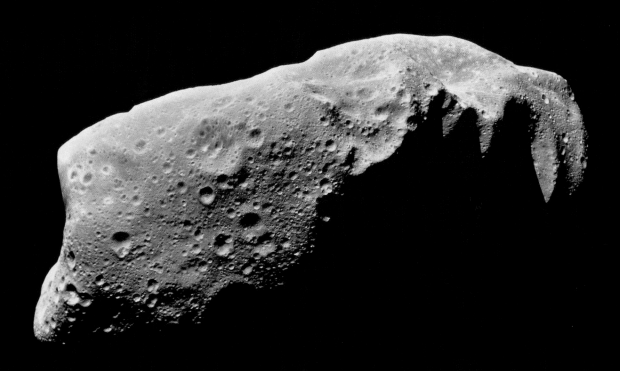

The planets were formed by millions of mountain-size, and smaller, chunks of rock and metal called planetesimals, which smashed into each other in the early life of the Solar System. The asteroids are leftover debris that was not swept up to become part of a planet. Pitted and battered asteroid Ida is three times the size of Manhattan.

EARTH AS A MOLTEN GLOBE

In its early days, Earth was a fiery-hot world glowing in the dark chaos of the disk material that still circled the young Sun. Each planetesimal that smashed into our youthful planet added heat, which kept Earth a soupy ball of molten rock and metals at a temperature of about 3600 degrees Fahrenheit (2000 °C). *Radioactive heating* from within also helped keep the planet molten. The main ingredients of the cosmic cloud out of which Earth formed were hydrogen, helium, carbon, nitrogen, oxygen, silicon, iron, nickel, aluminum, gold, uranium, sulfur, and phosphorus. Gradually, all of these materials began to separate from one another, the denser, or heavier, matter such as iron and nickel sinking into the core region while the lighter silicate materials floated to the surface and became crustal rock.

By about 3.9 billion years ago, a thin, solid crust of rock floated on a vast sea of molten rock. Gases bubbling out of volcano vents gathered together to form a primitive atmosphere. Meanwhile, water vapor, issuing as steam from countless geysers and from the impacts of cometlike chunks of ice from the solar disk, condensed as rain. The rains poured down for thou-

In the early 1900s, the German scientist Alfred Wegener imagined that some 250 million years ago an enormous supercontinent called Pangaea made up Earth's surface. Over time, Pangaea was broken apart by crustal movements into two lesser continents, one in the north called Laurasia and one in the south called Gondwana. By about 65 million years ago, the continents had broken apart even more into just about the positions they are in today.

sands of years, cooling the crustal rock as the water formed rivers and streams that flooded into basins and became the planet's first seas.

We can only guess what happened over the next hundreds of millions of years. But it now seems that by about 250 million years ago the planet's landmass was concentrated as a single supercontinent called *Pangaea*, which means "all Earth," and as a single ocean called *Panthalassa*, which means "all seas." Then about 200 million years ago powerful surges of magma from Earth's interior broke Pangaea into two lesser, but still enormous, landmasses. One was *Gondwana*, meaning "land of the Gonds," which refers to a people who lived in present-day India. It later separated into the present lands of Africa, South America, India, Australia, and Antarctica. The other, called *Laurasia*, fragmented to become Eurasia, North America, and Greenland. By about 65 million years ago, the general outlines of the continents closely resembled those of today. But that does not mean the continents had settled down for good. Forces from within the planet continue to move them to this day.

TWO

EARTH'S CRUST

We walk on a layer of rock called the crust that is as thin as an eggshell when compared with the rest of Earth. On average, the crust goes down only about 35 miles (56 kilometers) from the surface of the continents, and only about half as deep beneath the ocean floor.

To our eyes, the features of the crust seem permanent and timeless—mountains, valleys, vast underground caverns, deep trenches in the sea, and mighty rivers such as the Mississippi and the Amazon. But throughout geologic time the crustal features have formed and changed, and new features have replaced them. What was once land is now sea, and what used to be oceans are now lofty mountains thrust thousands of feet above sea level.

Many of the changes that have taken place in Earth's crust can be explained by plate tectonics. Earth's outer layer is made up of six larger, and a dozen or so smaller, plates—enormous rafts of rock. The largest plates include the Pacific Plate, the North American Plate, the Eurasian Plate, the

Antarctic Plate, and the African Plate. The plates grind against each other at their edges, pull apart, or smash together. Plate movement taking place now is gradually widening the Atlantic Ocean, shrinking the Pacific Ocean, and pushing the Himalayan Mountains even higher.

ROCKS AND THEIR MINERALS

Earth's rocky crust is made up of minerals mixed and packed together. Geologists have identified about 3,000 different kinds. Three common ones are quartz, feldspar, and mica. When they occur together, they form the type of rock known as *granite*. Some common minerals that are chemical elements include sulfur, copper, platinum, and silver. All of those minerals were once dust grains in the swirling solar disk out of which Earth was formed. Since that

distant time, the churning, melting, and pressing that have occurred within the planet's crust have rearranged its minerals to form three main types of rock.

Igneous rocks are born out of the fiery magma deep beneath the planet's crust. They include granite, the heavy dark rock called basalt, and volcanic lavas, among which is obsidian, also known as volcanic glass. The lava becomes obsidian when outpourings cool quickly and do not form crystals. When lava cools slowly it forms the fine crystals that are found in basalt rock. When magma solidifies slowly underground, the grains are much larger and it becomes granite. California's Sierra Nevada mountain range is made of granite, as is Georgia's famous Stone Mountain.

Stone Mountain, near Atlanta, Georgia, is a massive mound of light gray granite rock standing about 650 feet (200 meters) above the surrounding plateau. Igneous rock, almost free of major fractures, has resisted erosion by weathering and so continues to stand high.

Sedimentary rocks are formed out of sediments—the sand, gravel, mud, and silt carried by rivers and deposited in the oceans. Over time these sediments pile up in thick layers on the seafloor and are squeezed by the great weight above. Certain chemicals in the water meanwhile act as a glue that cements the soft sediments into hard rock. Millions of years later the sea may dry up and expose its sedimentary rock as layers of sandstone, shale, or a mixture of pebbles and small rocks called *conglomerate*. Limestone is a common sedimentary rock made of the compressed skeletons of trillions of tiny sea animals.

Metamorphic rock is the third main category of rock. Its name means "changed in form." It is created deep underground when pressure and high temperatures melt and resolidify other types of rock, turning them into completely different types by altering their mineral composition. Examples of metamorphic rocks are slate, marble, gneiss, and schist. Much of New England and eastern Canada is made up of this kind of rock.

To those who know how to read them, road cuts like this one in the state of Maryland are geologic history books. Layers of different rock types, squeezed and bent under tremendous pressure, formed one atop the other over millions of years as soft sediments of the seafloor were compressed into rock that was later thrust up out of the seafloor and exposed to view.

There is no beginning or end to a rock's physical evolution. In a continuous process called the rock cycle no new materials are created nor are any pre-existing materials lost. All are endlessly exchanged and recycled from one type of rock into another.

THE SEAFLOOR

If you could explore the seafloor you would find a world as foreign as the lava fields of Venus. Compared with the continental plates on which we live, the oceans cover almost three-quarters of the planet's total surface area of 197 million square miles (510 million square kilometers).

In the past, people imagined the ocean floors as flat and featureless, but now we know better. The mountains, cliffs, plains, and valleys on the land are minor compared with the rugged features that lie hidden in the ocean depths. Mountain ranges hundreds and thousands of miles long with peaks higher than Mt. Everest lie darkly in the Pacific deeps. When the peaks of these undersea mountains stick out of the water, they form islands. People who inhabit the Hawaiian Islands are actually living among the peaks of a group of volcanic mountains rising from the seafloor 6 miles (10 kilometers) below.

For the most part the features of the ocean floor alter little when compared with those on land. Rivers, sand, wind, and rain are constantly sculpting and altering the face of the land. But underwater the forces of erosion act gradually, so change is more slowly paced. But drastic changes do take place over very long periods of time. The crustal plates making up the seafloor collide or pull apart, and undersea volcanic outpourings build enormous mountain chains.

If you could walk all the way out to the seafloor, you would first cross a broad, flat area called the *continental shelf*, which stretches seaward for about 10 to 200 miles (16 to 320 kilometers), depending on where you happen to be. At the far edge of the shelf the water is from 200 to 600 feet (60 to 182 meters) deep. From there you would plunge down the *continental slope* for a distance of about 3 miles (5 kilometers) to the seafloor itself.

Sediments build up along the edge of the continental shelf. From time to time they break loose and cascade down the slope at 50 miles (80 kilometers) an hour. This creates *turbidity currents*, which are more destructive than the worst avalanche. These fierce currents of muddy sediments carve out enormous canyons as they rush down the slope. Several canyons stretch for 2 miles (3 kilometers), twice as deep as the Grand Canyon in northern Arizona. One, the Congo Canyon, is 500 miles (800 kilometers) long.

If you were crossing the floor of the Atlantic Ocean on your way to Europe, midway you would have to climb a 7,600-foot-high (2,315-meter-high) volcanic mountain range called the Mid-Atlantic Ridge. It extends for about 10,000 miles (16,000 kilometers) from the North Atlantic Ocean almost to Antarctica. Some of its peaks that poke out of the water include

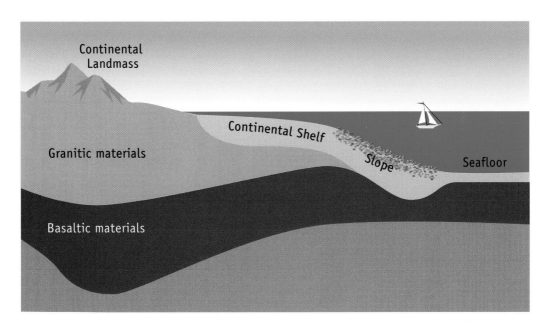

The continents can be thought of as lightweight granitic "rafts" floating in a "sea" of heavy basaltic rock. The continental shelf and slope are piled up with sand, clay, mud, and other sediments washed off the land.

21

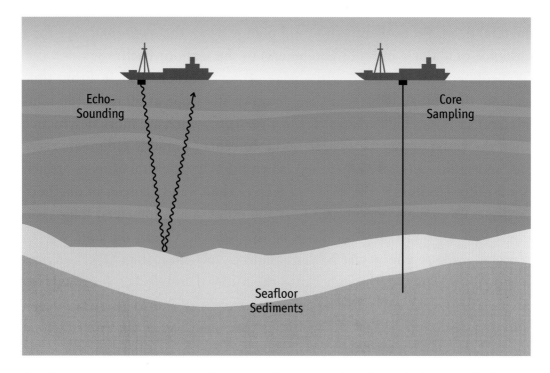

At left, a research vessel measues the depth of the ocean floor by exploding a small charge and then measuring the time it takes for the echo to bounce off the seafloor and return to the ship. Since the speed of sound through water is known, the depth of the bottom sediments can be calculated. At right a ship fires a hollow tube into the soft seafloor sediments. When the tube, now packed with sediments, is drawn up, the core of sediments can be removed and stored away for future study back in the laboratory.

Iceland, St. Peter and St. Paul Rocks, and the Azores. If you were strolling across the floor of the Pacific Ocean you would pass huge flat-topped formations, or seamounts, called guyots. They are 12,000-foot-high (3,660-meter-high) dead volcanoes whose tops were once above the waves but which, over time, were worn flat by the constant wash of waves. Today they are 5,000 feet (1,500 meters) beneath the ocean surface.

After the region of guyots, you might find yourself descending into a mammoth trench 6.6 miles (10.6 kilometers) deep. Called the Mariana Trench, it is located east of the Philippines and could swallow up Mt. Everest with room to spare. The trench is now being formed by one enormous crustal plate colliding with another and then being pushed down into the molten rock below. Wherever the plates have pulled apart and exposed deeper strata, the rock there is of the darker and heavier basalt type.

The floating continents are made of light-colored and lighter-weight granite-type rock. All the rocks that make up Earth's crust are part of what

Billions of skeletons of tiny sea creatures formed these limestone cliffs in New Mexico. Dead bodies of the tiny animals drifted to the seafloor, were buried and later compressed into the sedimentary rock type called limestone.

Strange structures formed by upwelling minerals from cracks in the seafloor, called hydrothermal vents, have been discovered in several regions of the seafloor in recent years. This carbonate tower has "grown" to a height of 33 feet (10 meters) and gone down to a depth of 24,500 feet (750 meters) in the North Atlantic Ocean off the Mid-Atlantic Ridge in an area named the "Lost City."

geologists call the *lithosphere*. Both the continents and ocean floors rest on a deeper puttylike layer of still heavier rock hundreds of miles thick and known as the *asthenosphere*. It begins about 60 miles (96 kilometers) beneath the surface and reaches a depth of about 450 miles (725 kilometers).

As we continue to explore the ocean floor, scientists realize more and more that it hides a storehouse of valuable information about Earth's long history. The sediments, piled up about 2,000 feet (610 meters) deep, are especially telling and important. For many years, scientists have driven hollow metal tubes down through them. Fossils, mineral-rich ooze, and other earth materials brought up in these tube *core* samples help to tell the story of Earth's mysterious past.

THREE

PROBING EARTH'S INTERIOR

Since scientists cannot drill a hole into Earth's core, they have had to use other, indirect ways of learning about the planet's deep interior. Their most important "tool" has been the *seismic waves* that earthquakes send shivering through Earth.

THE TRAVELS OF EARTHQUAKE WAVES

Whenever an earthquake bumps, jars, or shakes our planet, no matter how mild, pressure waves are sent out from the earthquake's focus, or point of origin, deep underground. One kind of wave recorded by earthquake instruments is the push-pull, or compression, wave. When the earthquake is triggered, the bump it causes in the rock is passed on to the neighboring rock. The chain then continues, stretching out to rocks farther and farther away, until the wave reaches Earth's surface.

Push waves are written as *P–waves*. The P stands for "primary" because this type of wave travels the fastest and reaches the earthquake station ahead of all other waves sent out by an earthquake. You can create your own compression waves by tapping the end of a steel bar with a hammer. The particles of metal struck by the hammer at the tip of the bar are first compressed but then expand again as they return to their normal shape. When they expand they compress the neighboring particles farther along until the hammer tap is passed the length of the bar.

The second kind of wave that reaches the earthquake station is a shake wave, also known as a shear wave. Because this wave travels slower than P–waves, it is called an *S–wave*, meaning secondary wave. You can set off a train of shake waves by striking a metal bar along its side rather than on the

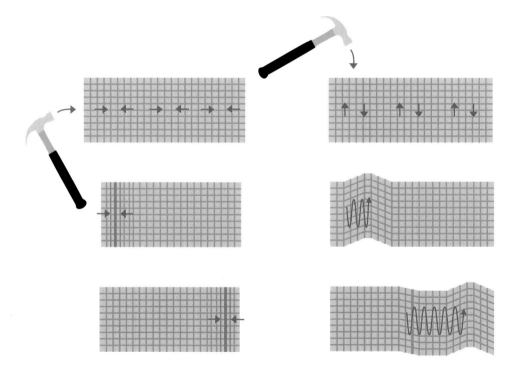

P–waves are set up if you strike the end of a metal rod with a hammer. These waves travel faster than S–waves and can travel through solids, liquids, and gases. S–waves are set up if you strike the top of a metal rod with a hammer. These waves are slower than P–waves and pass only through solids.

end. This produces an up-and-down motion in the bar's metal particles, which travels to the opposite end of the bar as a wave. You can probably guess what happens if you tap the bar with a glancing blow at the end so that you generate both P– and S–waves. Because the P–waves travel faster, the end of the rod will begin to vibrate back and forth a fraction of a second before the S–waves make it vibrate up and down.

During an earthquake both P– and S–waves leave the focus at the same time. But because the S–waves travel only two-thirds as fast as the P–waves, the P–waves are felt first. For instance, P–waves traveling through rock 50 miles (80 kilometers) deep travel at the rate of about 5 miles (8 kilometers) a second while S–waves trail behind at about 3 miles (5 kilometers) a second. Farther down, as the solid rock becomes denser, both P– and S–waves gather speed. In rock 1,700 miles (2,735 kilometers) deep, P–waves race along at 8.5 miles (13.7 kilometers) a second compared with S–waves' speed of 4.6 miles (7.4 kilometers) a second.

When P– and S–waves reach Earth's surface from their deep focus they generate a third kind of wave called *surface waves*. These longer waves travel slowest of all but are the ones that sometimes shake the ground hard enough to cause buildings to collapse. When earthquake scientists, called *seismologists*, found out more about the behavior of P–and S–waves as they

SEISMIC WAVE TRAVEL TIMES			
Region	Depth	P–Waves	S–Waves
	(miles)	(mi per sec)	(mi per sec)
Crust	0 to 25	speeds vary	speeds vary
Mantle	25 to 250	5.0 to 5.6	2.7 to 3.1
	250 to 600	5.6 to 7.1	3.1 to 4.0
	600 to 1700	7.1 to 8.5	4.0 to 4.6
	1700 to 1800	8.5	4.6
Outer core	1800 to 3100	5.0 to 6.8	–
Inner core	3200 to 3960	7.0 to 7.1	–

travel through rock, they were able to put the waves to work and so learn about the deep interior of our planet. One thing they discovered is that P–waves can travel through solid rock, through water and other liquids, as well as through the air. The reason is that the particles making up all three states of matter can be compressed. Sound waves can travel through all three for the same reason. S–waves, on the other hand, can travel through only those materials whose shape can be changed—in other words, only solids. The molecules of liquids and gases slip and slide around each other too so easily that they cannot transmit the up-and-down motion of shake waves. Knowing that P– and S–waves behave differently in various kinds of materials and travel at different speeds, seismologists were then able to "unlock" Earth's secret interior.

Discovering Earth's Strange Core

In 1906, the British geologist Richard Dixon Oldham announced that Earth had a core, a dense central region made of a material much different from the lightweight granite-type rock of the continents and the heavyweight basaltic rock of the seafloor. He was studying the path P–waves took as they traveled from earthquakes and moved through the planet. Imagine an earthquake with its focus near the North Pole, as shown in the chart on page 27. P–waves speed outward in all directions from the focus, but as they strike some denser material deep within Earth they are deflected slightly. The result is a shadow zone at the surface nearly empty of P–waves. To Oldham this indicated that the planet had a core of material that seemed to be much denser, or heavier, than either the continents or the seafloor.

Later, other investigators found that S–waves seemed not to be passing through the newly discovered core at all. Could the core be made of some kind of very dense liquid? The answer seemed to be yes, since S–waves can't travel through liquids. The young Earth's original supply of molten iron must have flowed and settled, collecting as an iron ball inside a rock-coated ball in the heart of the planet. And due to the great pressure in the core,

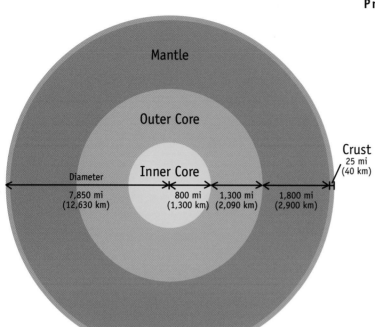

Mantle

Outer Core

Inner Core

Crust
25 mi
(40 km)

Diameter

7,850 mi
(12,630 km)

800 mi
(1,300 km)

1,300 mi
(2,090 km)

1,800 mi
(2,900 km)

EARTH'S STRUCTURE

Geologists who study Earth's interior have discovered zones of different kinds of materials. At the surface is a thin layer of lightweight crustal rock. Deeper down is an enormous zone of much denser rock—the mantle. The planet's central region consists of an outer core of liquid iron-nickel within which is a smaller core of solid iron-nickel. (Radius distances shown in this diagram are approximate only.)

the iron remained molten instead of turning solid. In 1913, seven years after Oldham's discovery, the German-American seismologist Beno Gutenberg was able to show that the boundary of the liquid metal core was 1,800 miles (2,900 kilometers) below the surface. It appeared, then, to make up most of Earth's volume and two-thirds of its mass. The current belief is that the core makes up 16.4 percent of Earth's volume and one-third of its mass.

Another important boundary inside Earth had been discovered in 1909 by the Croatian seismologist A. Mohorovičić. While analyzing the wave patterns produced by a Balkan earthquake, he found that at a depth of about 37 miles (60 kilometers) the speed of P– and S–waves suddenly increased and continued to do so until they reached the core. By this time it was known that seismic waves gained speed as the rock through which they traveled grew denser. So Mohorovičić's findings indicated that there must be some sort of a boundary zone between the bottom of the crust and the top of the mantle and that its material had to be considerably denser than the crust.

The zone was named after Mohorovičić and is known as the *Moho*. It seems to be made up of some iron-magnesium silicate. The mantle pressure is so high that the material, although above its melting point at sea level, behaves like a solid.

A Core Within a Core

But the story of Earth's interior didn't end there. In the early 1930s, a Danish seismologist, Inge Lehmann, began to study the few P–waves that sometimes managed to enter the shadow zone. She wondered how they were able to get through. In 1936, she concluded that within the molten core there must be a smaller solid core that speeded up the P–waves and bent them so sharply that they skipped right into the shadow zone. Others agreed with her finding, and eventually it was determined that the size of the inner solid core was about 1,600 miles (2,575 kilometers) in diameter, which put its outer boundary about 3,100 miles (5,000 kilometers) beneath Earth's surface. The picture of the deep interior seemed fairly complete— one metallic core inside another, both made up mostly of iron with some nickel mixed in.

Considering that we cannot scoop out samples of Earth's interior, it is not surprising that the planet's core has been so long in giving up its secrets. But the strides made in the last century have been considerable. In the future, scientists are bound to make fresh discoveries as they develop new technologies to substitute for that deep hole we'll never be able to dig. In the meantime, seismologists are letting earthquake waves do the digging for them.

FOUR

EARTHQUAKES

The time was one minute twenty-eight seconds before noon, September 1, 1923. The place was Tokyo, Japan. The day had begun much like any other. At the noon hour, women shoppers, delivery boys, businessmen, and schoolchildren were preparing for lunch. There was no sign, no warning to indicate that the greatest earthquake disaster in history was about to crumble the cities of Tokyo and Yokohama.

Then it happened. The rock floor beneath the waters of Sagami Bay snapped. Sharp vibrations fanned out to Yokohama some 50 miles (80 kilometers) away and then quivered on to Tokyo some 70 miles (113 kilometers) away. The most violent tremors shook the ground for about thirty seconds. Then came a series of rapid, less severe aftershocks lasting several minutes.

This is how Professor Akitune Imamura, a seismologist at Tokyo University, described the scene:

> At first, the movement was rather slow and feeble, so I did not take it to be the forerunner of so big a shock. . . . Soon the vibration became large, and after three or four seconds from the commencement, I felt the shock to be very strong. Seven or eight seconds passed and the building was shaking to an extraordinary extent, but I considered these movements not yet to be the principal portion. At the twelfth second from the start came a very big vibration, which I took at once to be the beginning of the principal portion. Now the motion, instead of becoming less and less as usual, went on increasing in intensity very quickly, and after four or five seconds I felt it to have reached its strongest. During this time the

A crumbled building shows the destructive force of earthquake surface waves. When P– and S–waves that are generated at the earthquake's epicenter reach the surface, they set up the long, slow, and destructive surface waves.

tiles were showering down from the roof making a loud noise, and I wondered whether the building could stand or not. During the following ten seconds the motion, though still violent, became somewhat less severe, the vibrations becoming slower but bigger. For the next few minutes we felt an undulatory movement like that on a boat.

Soon after the first shock, fire broke out at two places in the university, and within one and a half hours our Institute was enveloped in raging smoke and heat. It was ten o'clock at night before I found our Institute and Observatory quite safe. We all did our best, partly in continuing earthquake observations and partly in extinguishing the fire, taking no food or drink till midnight.

The section of Tokyo built on soft soil and landfill suffered the worst damage from the quake itself. Fifty-four percent of the city's brick buildings and 10 percent of its reinforced concrete buildings were destroyed or badly damaged. Of sixteen steel frame buildings, only six were unharmed. The shifting and shaking ground throughout the city bent and snapped water lines. So by the time the hundreds of fires that had broken out reached full force and swept across the city, the fire department was helpless. Within half an hour after the earthquake began, 136 fires raged in Tokyo alone. After 56 hours a total of 366,262 houses were gone—almost 75 percent of Tokyo's homes. The Yokohama fire was even more severe. Within 12 hours, more than half the city was in total ruin, and by the time the last fires died out the entire city was gutted.

The death toll for the areas affected by the earthquake tremors and then destroyed by fire reached about 140,000 people. An additional 100,000 were injured and about 43,000 were missing. Half a million were left homeless. In Tokyo during the remainder of September seismographs recorded a total of 1,700 aftershocks.

FAULTS

During the Japan earthquake of 1923 a great *fault* along the bottom of Sagami Bay snapped. Thousands of faults thread their way over the globe. A fault is a significant crack in Earth's rock crust, the surfaces of which can slip back and forth or up and down against one another. Year after year the two surfaces of a fault may press against each other without causing a disturbance. But after a while the pressure becomes so great that the two surfaces eventually slip, then snap, as when you snap your fingers. When they do, violent vibrations race through the ground and may wreak havoc for miles around.

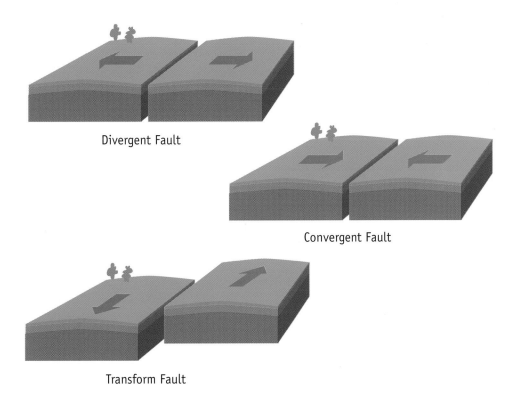

Divergent Fault

Convergent Fault

Transform Fault

An earthquake fault line occurs where the edges of two plates of Earth's crustal rock meet and slip against one another. Divergent faults move away from one another. Convergent faults ram into one another. Transform faults slide past each other.

One of the most spectacular, visible faults is the San Andreas Fault, located in California. This break in Earth's crust extends for some 650 miles (1,050 kilometers) across California's coastal ranges of mountains. In 1800, the fault was at rest. But by 1875 the ground along the fault had become warped by the decades during which its two surfaces had been gradually straining against one another. Then on the morning of April 18, 1906, the fault snapped along a 270-mile (435-kilometer) section that runs from Upper Mattole in Humboldt County to San Juan in San Benito County. The energy suddenly released had been built up and stored, as is the energy in a stick that you slowly bend. When bent too far, suddenly the stick snaps and releases all of that stored energy as a "stickquake." When the San Andreas fault snapped, the ground along it shifted horizontally, producing some bizarre effects. A road that crossed part of the fault line was broken in two; one part of it displaced to the right, the other to the left. The front walk of a ranch house hopped several feet in one direction while the house shifted in the opposite direction.

San Francisco, only 10 miles (16 kilometers) away from the fault and the largest city in the area where the earthquake struck, was hard hit when the rock shifted sharply. It took only a minute to destroy the new City Hall building. Other buildings with their foundations in soft soil were twisted and cracked, but those standing on rock foundations escaped with little initial damage. As with the Tokyo earthquake, it was the fire that followed that accounted for about 95 percent of the damage. Like other faults, the San Andreas may jolt suddenly, or it may release its stored energy in a slow, steady motion called a creep. The fault is at the border of the North American Plate and the Pacific Plate, which are slowly grinding past each other in a north-south direction. One of California's costliest earthquakes struck near the southern California city of Northridge on January 17, 1994, killing more than sixty people and causing an estimated $30 billion in damage.

California's San Andreas Fault is a crack in Earth's crust where the North American Plate rubs up against the Pacific Plate and, from time to time, triggers earthquakes. This occurs when plate stresses are released and the ground jumps.

TSUNAMIS

When faults snap on land, their destructive force is limited to the vibrations they send out. But when they snap along the seafloor they may set up "tidal waves," which seismologists call *tsunamis*, since these powerful disturbances have nothing to do with the tide. When an undersea fault snaps up and down, rather than from side to side, huge amounts of energy are transferred to the water and build a series of broad low waves that travel at speeds of up to 500 miles (800 kilometers) an hour. Ships at sea do not notice a train of tsunamis since each wave may be only about 2 feet (0.6 meter) high, and the distance between individual waves can be 100 miles (160 kilometers) or more.

The full force of a major tsunami isn't felt until it sweeps into shallow water and crashes onto the shore. As it approaches the land, it begins to drag against the bottom. That causes a rapid decrease in speed to only about 50 miles (80 kilometers) an hour. But what it loses in velocity it quickly gains in height. The wave towers to 100 feet (30 meters) or more as the water behind it piles up. Finally, the tsunami crashes down with terrible destructive force.

On June 15, 1896, an undersea earthquake 125 miles (200 kilometers) from the Sanriku district of Japan set up a tsunami that swept in and reared to more than 100 feet (30 meters) before it pounded onto the shore. After the waters receded, 27,000 bodies and debris from more than 10,000 houses lay strewn about or were carried out to sea by the backwash of the wave.

Faults are not the only causes of earthquakes. Some seismologists suspect that large volumes of magma deep under the crust sometimes shift about with explosive force. Such disturbances not only send P– and S–waves quivering through the planet but may also create new faults. Some Japanese seismologists have reported that the level of certain areas of land has been either raised or lowered as a result of an earthquake. This may be evidence of pressure changes within deep pools of magma.

EARTH AS A MAGNET

Our planet is a giant magnet with lines of force looping from the North Pole to the South Pole. Like the causes of earthquakes and volcanoes, the secrets of Earth's magnetism are buried deep in the interior. For more than a thousand years sailors have used Earth's magnetism to steer their ships. The earliest compasses may have been nothing more than a magnetic rock, called magnetite, resting on a piece of wood floating in a dish of water. As one compass user of ancient times wrote: "A stone so placed in a boat will turn until the north pole of the stone will come to rest in the direction of the North Pole of the heavens. And if you move the stone from that position a thousand times, a thousand times will it return by the will of God." Although sailors of old knew that Earth was a magnet, they had no idea why.

Nails pop out of this wooden ship, causing it to sink and its sailors to drown in this sixteenth-century woodcut. This illustrates a belief at the time that strongly magnetic islands could cause such calamities.

Scientists now think that the iron-nickel core may act like a dynamo, a machine that can change mechanical energy into electrical energy. The mechanical energy is the continuous movement of the core's metallic mix. The motion, possibly driven by intense heat or by chemical differences within the core material, generates electrical currents and magnetism. Earth's rotation also comes into play and somehow lines up the local magnetic effects to produce the planet's looping magnetic field. But it is not entirely clear whether this is what actually happens.

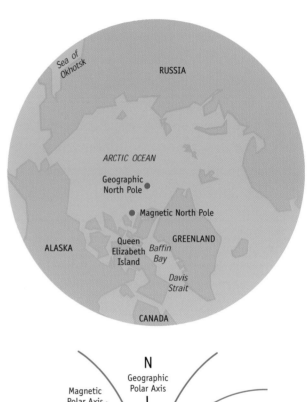

Earth is a giant magnet with magnetic lines of force looping around from the Magnetic North Pole down to the planet's Magnetic South Pole. Churning within Earth's liquid outer core is a mix of iron-nickel thought to generate Earth's magnetism. While the planet's geographic North and South Poles remain fixed, the magnetic poles wander around. The diagrams show the present position of the Magnetic North Pole to the left of Greenland and the path the shifting pole has followed from 1831 to 1994. At present, the pole moves at the rate of 9 miles (15 kilometers) a year.

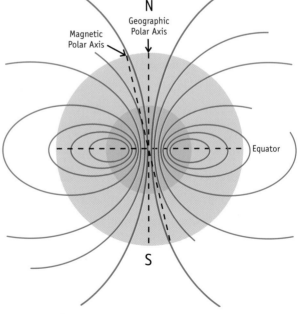

One thing that is certain is that the planet's magnetic poles have been in motion through Earth's history, but again no one is exactly sure why. The poles have even flip-flopped many times, or reversed positions. Today the Magnetic North Pole is between Bathurst and Prince of Wales Islands in the Canadian Arctic. That puts it about 800 miles (1,300 kilometers) from the Geographical North Pole. And the Magnetic South Pole is located off the coast of Wilkes Land, Antarctica, about 1,600 miles (2,550 kilometers) from the Geographical South Pole.

The Magnetic North Pole got to its present position by drifting thousands of miles along a looped path over millions of years. For instance, about 100 million years ago it seems to have been located in Siberia; 345 million years ago it was in China; 430 million years ago it was northeast of the Philippines; and about 550 million years ago it was in the central Pacific. Evidence for the wandering poles are the alignments of magnetic grains found in rocks from various periods in the planet's geological past and from different parts of the world. When the rocks originally cooled, their minerals stayed oriented in the magnetic field existing at that time.

Although it may sound as though the poles themselves are doing the moving, that is not the case. The poles actually remain fixed. Instead, it is Earth's crust that is shifting its position relative to the poles.

FIVE

VOLCANOES

On May 8, 1902, a few minutes before eight in the morning, 40,000 people were engulfed by an immense cloud of flame that swept down the slopes of Mt. Pelée, a volcanic mountain overlooking the city of St. Pierre on the island of Martinique in the West Indies. In a matter of minutes what had once been a picturesque coastal city was reduced to charred ruins by the fury of an exploding volcano.

About two weeks before the eruption residents of St. Pierre had noticed wisps of smoke rising lazily from the quiet mountain. Hikers climbed the slopes to investigate and returned saying that they had heard rumblings deep within the mountain. Pelée, they feared, was about to go on a rampage. But others were skeptical, saying that Pelée had been sleeping for fifty-one years. The possibility of its erupting seemed remote.

A raging cloud of flame and ash from the 1902 eruption of Mount Pelée rushed down the mountain and totally destroyed the town of St. Pierre on the island of Martinique. About 40,000 people were killed.

As the days wore on the townspeople began to hear muffled explosions within the mountain. Then peaceful Pelée awoke. It began belching clouds of ash from its crater. Within a few days the island was coated with a thin film of white ash. Instinctively animals moved down from the slopes, away from the disturbance. On May 5 there were more explosions, this time followed by sprays of boiling mud and an artillery of rocks fired from the crater. Many people tried to leave St. Pierre, but the local governor reportedly posted soldiers around the town to prevent anyone from leaving; elections were only a few days away. Then on that fateful morning three days later, Pelée, like a giant flame thrower, hurled its breath of fire down the mountainside onto St. Pierre. All were killed except one person—a prisoner in the local jail. He was locked in an underground cell, and so was protected from the explosion.

Of the more than one dozen ships anchored in St. Pierre's harbor, all but one caught fire and were destroyed. Assistant Purser Thompson of the steamship *Roraima* was an eyewitness to Pelée's outburst.

> As we approached St. Pierre we could distinguish the rolling and leaping of the red flames that belched from the mountain in huge volumes and gushed high into the sky. Enormous clouds of black smoke hung over the volcano. . . . The flames were then spurting straight up in the air, now and then waving to one side or the other a moment, and again leaping suddenly higher up.
>
> There was a constant muffled roar. There was a tremendous explosion about 7:45 o'clock, soon after we got in. The mountain was blown to pieces. There was no warning. The side of the mountain was ripped out, and there was hurled straight toward us a solid wall of flame. It sounded like thousands of cannon. The wave of fire was on us and over us like a lightning flash. It was like a hurricane of fire . . . The fire rolled [in mass] straight down upon St. Pierre and the shipping.

The town vanished before our eyes, and the air grew stifling hot and we were in the thick of it.

Wherever the mass of fire struck the sea the water boiled and sent up vast clouds of steam. I saved my life by running to my stateroom and burying myself in the bedding. The blast of fire from the volcano lasted only a few minutes. It shriveled and set fire to everything it touched. Thousands of casks of rum were stored in St. Pierre, and these were exploded by the terrific heat. The burning rum ran in streams down every street and out to the sea. . . . Before the volcano burst, the landings of St. Pierre were crowded with people. After the explosion not one living being was seen on land.

The fury of an erupting volcano often claims the lives of those unfortunate enough to be in the vicinity. In the year A.D. 79 Mount Vesuvius in southern Italy erupted after three hundred years of sleep. Indications that the mountain was awakening came seventeen years earlier in the form of violent earthquakes. Then around noon on August 24, the mountain burst forth in a succession of terrible explosions. First came suffocating clouds of dust and hot gases. Then white pumice rained down and completely buried the town of Pompeii to the southeast. Next stony matter, soaked with water and fused together, welled up within the mountain, overflowed the crater rim, and rushed down the slopes in a mudflow that buried the town of Herculaneum to the southwest. Hundreds of families were buried alive by seething lava, which cemented them in place as they fled and the whole city was covered by 13 feet (4 meters) of ash.

The Roman scholar Pliny the Younger witnessed the eruption and wrote a detailed account of it. Not until 1709, seventeen centuries after the eruption, when the remains of devastated Pompeii were first excavated, could people fully comprehend the magnitude of what happened on that tragic day in A.D. 79.

Smothered and suffocated where they fell, victims of the A.D. 79 eruption of Italy's Mount Vesuvius were instantly buried under ash from the volcano, as was the entire city of Pompeii. Archaeologists later excavated the ancient city and found grizzly human remains like those shown above. Below that is the unearthed city, as it now appears.

The most powerful volcanic eruption in history was that of the Tambora Volcano in Indonesia in 1815. It spewed out a cloud of sulfur dioxide and ash that dimmed the Sun for a year and lowered global temperatures by as much as 5 degrees Fahrenheit (3 °C). In parts of Europe and North America, 1816 was known as the "year without a summer." Since the year 1600, nearly 300,000 people have been killed by volcanic eruptions.

ANATOMY OF AN ERUPTION

The world's largest volcano is Mauna Loa, part of the island of Hawaii. It towers nearly 6 miles (10 kilometers) from its base on the ocean bottom. It is one of about fifteen hundred major active volcanoes in the world today, not including many yet undiscovered on the seafloor. Volcanoes give us a direct way of examining the kinds and the nature of the materials located beneath Earth's crust. With violent volcanic eruptions we can expect a certain sequence of events. First there are small, local earthquakes surrounding the mountain about to become active. The earthquakes may be accompanied by deep-throated rumblings. Sometimes a lake near the volcano suddenly disappears or changes level just before the mountain erupts as the unstable ground supporting the lake cracks open or rises. This usually happens just before the actual eruption. Great blasts of steam, sometimes thousands of feet high, then begin to roar out of the crater.

Mixed with the steam are gases, rocks, dust, and ash, which are tossed and blown about. When the famous volcanic island of Krakatau, near Java and Sumatra, blew up on August 26, 1883, it erupted in a series of explosions. Then, on the following day there was one mighty blast of flame, smoke, and ash that rose 17 miles (27 kilometers) into the air. Krakatau was uninhabited, but thousands of people on Java and Sumatra were killed by waves more than 115 feet (35 meters) high caused by the eruption.

When the sea calmed and the air cleared, there was nothing left. The mountain was gone. People 3,000 miles (4,800 kilometers) away in Madagascar heard the explosion. A series of gigantic waves generated by the

blast washed away 165 coastal villages on Java and Sumatra, killing 36,000 people.

The steam rising out of an erupting volcano condenses and falls back to the ground as rain. As it falls, it mixes with the dust and ash and splatters to the ground in torrents of mud. Thunder and lightning boom and flash around the top of the mountains. Eventually, magma deep within the mountain wells up and floods over the crater walls, pouring down the mountainside in great glowing rivers of lava.

Meanwhile, blobs of glowing lava may be shot out of the crater and hurled thousands of feet into the air. These volcanic bombs solidify and fall back to the ground. Cotopaxi in Ecuador reportedly tossed a 200-ton block of hardened lava 9 miles (14 kilometers).

As the gases escape, the magma becomes more fluid, welling up to the surface more rapidly. Magma pours out of a volcano as red-hot or white-hot lava at a temperature of about 2000 degrees Fahrenheit (1100 °C). Most magmas originate in the lower crust and upper mantle at a depth of about 100 miles (160 kilometers). Very thick, gassy, and explosive magma issues from volcanoes on land as compared with the less active, more gently flowing magma that pours from undersea volcanoes. Magma with lots of silica, typical of crustal rock, is usually more violent because it is more gassy. The main thing that determines the nature of the magma a volcano spews is the kinds of rock it melts through on its way up from the mantle and through the crust. Magma may take from thousands to millions of years to make that journey.

When forces within Earth shift the rock that lies along a volcanic fault line, pressure on the hot basalt below is often reduced. This allows the basalt to melt and flow upward through the fault zone as magma. Tongues of magma leading the main flow may seep through cracks in rock layers

The craters of many volcanoes have lakes of highly acidic water, like this one in Irazu National Park, Costa Rica. One such crater lake in Kamchatka is so strongly acidic that a nail dropped into it dissolves in two days.

Sometimes the lava spewed out of a volcano is so thick that it is like toothpaste squeezed out of a tube. The lava is so close to being solid that it freezes into a towerlike shape. California's Lassen Peak is one example.

and become features called *dikes*. Or the magma may force its way between pancake layers of rock and stay there to form intrusions called *sills*. As the magma flows toward the surface the pressure exerted on it becomes less and less. This allows gases trapped in the magma to bubble out—as carbon dioxide bubbles out of a can of soda when you reduce the pressure by opening it. The greater the rate of escape of the magma's gas, the more violent the eruption.

The gases released during an eruption are mostly water vapor and carbon dioxide, along with some sulfur dioxide, hydrogen sulfide, and hydrogen. The gas cloud billowing out of the crater is more than 90 percent water vapor. Scientists are not sure just where all this water comes from. Possibly it can be traced to oxygen combining with hydrogen in the magma, groundwater locked in the rocks, or maybe both.

One of the most recent theories accounting for volcanoes is the *plume theory*. Hot rock in parts of the lower mantle near the edge of the outer core rises into the upper mantle in the shape of a giant mushroom with a cap a few hundred miles wide. The magma cap then melts its way up to the surface as a "hot spot" eruption of lava. Outside the region of a plume's activity an area of cooler mantle rock may be drawn down to replace the rising hot rock. In this way circu-

The May 1980 eruption of Mount Saint Helens in Washington State is a reminder that Earth's active interior can trigger violent eruptions of volcanoes at a moment's notice.

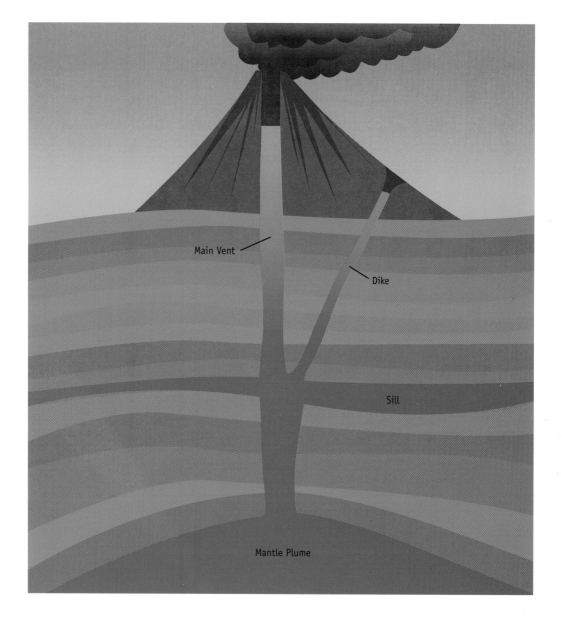

lating pockets of magma may keep parts of the mantle stirred. To date, scientists have identified about twenty major hot-spot plumes around the world, including the Hawaiian Islands and Yellowstone National Park. As with most of Earth's deeply rooted phenomena, our knowledge is growing slowly.

SIX

MOUNTAINS

Time stands still in the mountains. Or so it seems. Although mountains appear to be permanent fixtures, this is only an illusion. A rock-solid mountain is as sure to crumble, in time, as a sand castle. Earth has seen about ten major periods of mountain building. During those times the Alps of Europe, Asia's Himalayas, and Russia's Urals, to name a few, were raised. But ultimately these mighty formations are worn away by erosion as new ones are made. Some 225 million years ago the young Appalachians, now gentle and rounded with age, were splendid snow-capped peaks, possibly as striking as the Alps. They ran from Newfoundland to Alabama. But over thousands and hundreds of thousands of years the forces of erosion have smoothed and worn them down.

How Do Mountains Form?

We are now living in an active period of mountain building. In fact, ever since people have roamed this planet, mountains have been buckling up from its crust. The Andes of South America, our own Rockies, and the Himalayas are all young mountain chains that have been thrust up within the past 60 million years or so. The dinosaurs never saw them.

Scientists understand the birth and formation of volcanic mountains far better than the development of mountains that are folded out of Earth's crust. One reason is that volcanic mountains are sometimes born overnight and reach a height of several hundred feet in a matter of days or weeks. Scientists can study the formation of these "laboratory" mountains more easily, due to their rapid growth spurts. But mountains such as the Alps, which are not volcanic, take millions of years to form.

Although scientists still don't know everything about how mountains are born, the theory of plate tectonics offers the best explanation so far. Accordingly, most mountains are formed by the movement of large plates in Earth's crust colliding with each other. But other forces, such as heat

Among the most spectacular recently formed mountain ranges are the Himalayas in Nepal. Appearing solid and tall, these grand, sharp peaks will one day be worn and crumbled by weathering and time.

currents in the upper mantle rock, as well as chemical and physical changes in the materials of the crust and mantle, also play a role in mountain building.

When you look at a mountain, there is much more than meets the eye. Like icebergs, which stick only their heads out of the water, mountains are thought to poke only their peaks above the ground. The bulk of their mass, called a mountain's roots, is housed deep in the crustal rock. Because geologists cannot see or collect samples there, they cannot be sure about the nature of a mountain's roots.

Even so, they do have an explanation for how a mountain is worn down. Erosion may shave a 10,000-foot (3,000-meter) peak down at the rate of two inches (five centimeters) a thousand years. In some 60 million years, the

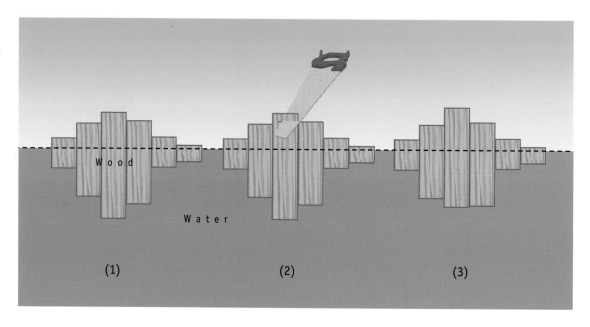

At (1) six blocks of wood of different weight float at different heights. At (2) if part of the top of the heaviest block is cut off, it now weighs less. At (3) since the cut-off block is now lighter, the pressure of the water below pushes the block higher. A mountain top "cut off" by erosion also becomes lighter in weight, and the upward lifting pressure acting on the mountain's roots tends to keep the mountain at just about the same height above the surrounding ground, just as the wood block at (3) is kept at the same height above the water surface.

mountain will be worn flat. But mountains also have a way of recovering their lost height.

We can regard a mountain as much like an iceberg, a huge rock floating in a sea of heavier, fluid rock. Over the centuries the mountain's hard rock is worn away and washed down onto the surrounding lowland as sediments. As the mountain loses weight by erosion, the surrounding lowland gains weight by collecting the eroded materials. The increased weight of sediments pressing down on the lowlands forces some of the fluid material below to move to a region of less pressure. That region is under the mountain, whose mass is slowly reduced. As more and more fluid rock collects under the mountain's roots, it pushes the mountain up. The faster the rate of erosion, the faster the mountain rises, while more or less retaining its height. This process is called *isostasy*, which means "equal standing."

KINDS OF MOUNTAINS

Folded mountains form the world's largest mountain systems. They are made of thick sedimentary rocks that long ago were thrust up out of the crust, which squeezed weaker sedimentary materials into folds. Switzerland's Jura Mountains and the Appalachians, and the Rockies in North America are all folded mountains.

Fault-block mountains are formed when great breaks occur in Earth's crust. When this happens, the ground on one side of the break may slip down hundreds of feet, while the ground on the other side remains as it was. The result is a solid rock wall that may be hundreds of feet high. When the slippage is great enough, a block mountain—also called a fault-block mountain—is formed. Eventually, rain and wind erode the sharp edges of the block, leaving the mountain with a rounded top. Utah's Wasatch Range, California's Sierra Nevada mountains, and the Teton Range in Wyoming are all fault-block mountains.

Folded mountain ranges, like these of Mount Cascade in Canada's Banff National Park in Alberta, are formed when weak sedimentary rock is thrust up by the wrinkling of Earth's crust. The folding process leaves two kinds of features—up-arches called anticlines, and down-arches called synclines. Switzerland's Juras and the United States's Appalachian Mountains are examples of the anticline-syncline type of system.

Volcanic mountains are formed when molten rock flows up through vents in Earth's crust and solidifies as lava. Chains of mountains can be formed this way. Italy's Apennines and Alaska's Aleutian Islands are examples of volcanic mountain chains. Examples of isolated volcanic mountains are Japan's Mount Fuji, Mount Shasta in California, and Mount Hood in Oregon. Outpourings of lava, cinders, ash, and other materials through a pipelike opening in the crust pile up over the years and build these lone cone-shaped formations.

Residual mountains result from large, flat areas of land rising above the surrounding plain rather than folding upward. This creates a high plateau. After many years, water draining off the plateau cuts into and eats away the softer earth and rock. At first only shallow ridges of hard rock are left exposed, but over the years these ridges tower higher and higher as the softer parts of the plateau are eroded away. The Allegheny Plateau of Pennsylvania is one such region that features residual mountains.

Fault-block mountains are formed when great blocks of Earth's crust break apart and tilt over so that high, flat cliffs are left exposed. Millions of years of erosion round out the sharp edges of blocks so they end up looking like these fault blocks in Lake Mead, Nevada.

Earth's history has consisted so far of about ten major periods of mountain building. Younger ranges like the Rocky Mountains of the Western Hemisphere, about 75 million years old, still poke their sharp peaks skyward.

GLACIERS

Over the past few million years, Earth has played host to several glacial periods, when vast sheets of thick ice covered much of the planet. Over the past 700,000 years there have been seven known glacial episodes. Each was then separated by a period of warming. Those of us presently living in the Northern Hemisphere may be enjoying the peak of one of these warm interglacial periods now.

Glacial periods work in cycles. Each cycle—from the peak of one glacial period, through an interglacial period, then to the peak of the next glacial period—lasts about 100,000 years. From the end of one glacial period to the beginning of the next is a span of from about 10,000 to 20,000 years. Just why ages of ice come and go is not fully understood. Some scientists think changes in the Sun's energy output may be responsible. Other theories suggest slight changes in Earth's orbit that move us closer to or farther away from the Sun from time to time. Perhaps periods of mountain building, or cooling due to the Solar System passing through clouds of cosmic gas and dust as the Sun moves through space also cause these changes. A dusty Solar System would mean that less solar energy would reach Earth until the dust cleared.

The last glacial period ended about 10,000 years ago. At its peak roughly 18,000 years ago, ice covered approximately 30 percent of Earth's total land surface. The ice probably first formed at higher altitudes as mountain glaciers and then grew into sprawling ice sheets that gradually crept south, smothering the land beneath a 2-mile (3-kilometer) thickness of ice.

There is more ice around the world today than you may think. Small sheets of ice and glaciers filling deep valleys are on every continent except Australia. Some of the world's most spectacular glaciers surround the Gulf

Rivers of glacial ice, like Alaska's Gilkey Glacier, are reminders of times when ice covered much of the planet to depths as great as 2 miles (3 kilometers). Ice ages come and go in cycles, the most recent one ending about 10,000 years ago.

of Alaska. Africa's Mount Kenya and Mount Kilimanjaro also have impressive specimens, as do the Pacific islands of New Guinea and New Zealand. More than one thousand glaciers decorate Europe's Alps while Europe's Caucasus Mountains are sprinkled with hundreds of these sprawling blocks of ice. Some are small, only the size of a football field. Others such as the great Aletsch Glacier in the Swiss Alps, stretch away for many miles. The Fedchenko Glacier in Asia's Pamirs range is one of the world's largest ice cubes—3,000 feet (915 meters) thick and 48 miles (77 kilometers) long.

SOLVING EARTH'S PUZZLES

The forces and processes at work within Earth will continue to change our planet as they have over the past 4.6 billion years. Understanding how they combine to bring about change is often a puzzle whose pieces keep changing shape. For instance, we mostly understand how glaciers form, move over the land, and change the surface as they move. But what causes ice ages to come and go is a piece of the puzzle that does not yet fit and whose shape can be expected to change as our knowledge about ice ages grows. Although the picture has become fuller, it is not yet complete.

We do not have one sweeping tell-all theory to explain how Earth as a planet works, just as there is no single, grand theory to explain how the Universe works. But that doesn't discourage scientists from searching for answers, and it never will. As geologist William A. Bassett of Cornell University puts it: "It is the excitement of finding the missing pieces and seeing a picture of some place that is so utterly inaccessible slowly emerge that makes this kind of research so exciting and rewarding."

GLOSSARY

asthenosphere The zone within the upper mantle where the rock can be squeezed and permit movement of the crust. It begins about 60 miles (100 kilometers) down and extends to a depth of about 450 miles (700 kilometers).

composition The elements of a substance.

conglomerate A rock type that is a mixture of sand, pebbles, and larger rocks.

continental shelf The broad and relatively flat, gently sloping platform of a continent that extends into the sea from a few miles to 100 miles wide (160 kilometers) or more.

continental slope The sloping outer edge of a continental shelf.

core The central region of Earth composed of iron and nickel and consisting of a solid inner core and a liquid outer core. The entire core region begins at a depth of about 1,800 miles (2,900 kilometers) from Earth's crust to about 4,000 miles (6,400 kilometers) at Earth's center.

crust The outermost layer of Earth's surface. The continental crust is some 25 miles (40 kilometers) thick. The ocean floor crust is about 3 miles (5 kilometers) thick.

dike A long, wide, but thin body of intrusive igneous rock that cuts across the layering of the surrounding rock formations.

dynamo A rotating machine that changes mechanical energy into electrical energy; it can also do the reverse.

fault Any break in Earth's crustal rock.

fault-block mountains Mountains formed when great breaks occur in Earth's crust.

folded mountains Mountains made of thick sedimentary rocks that long ago were thrust up out of the crust, which squeezed weaker sedimentary materials into folds.

Gondwana The southern half of Pangaea.

granite An intrusive igneous rock type with large grains made of quartz, feldspar, and mica.

guyot Flat-topped structures rising out of the floor of the Pacific Ocean, probably volcanic in origin, and the peaks of which have been worn flat by wave erosion.

igneous rock Rock formed when molten material flows up from deeper parts of Earth's crust and solidifies either within the crust or at the surface.

isostasy A condition of balance in which a large and high landmass is supported not by solid rock beneath, but by floating in the denser but fluid rock of the underlying mantle.

Laurasia The northern half of Pangaea.

lava Magma after it has been released by a volcanic eruption.

lithosphere Earth's rigid outer rock layer—the crust—that lies above the asthenosphere.

magma Fluid rock material originating in the deeper parts of Earth's crust and below.

mantle The layer of rock beneath Earth's crust, basically made up of iron and magnesium mixed with silicates. Because the mantle rock is under great pressure from the weight of rock above, the upper mantle is hot and behaves more like putty than a solid. Lower down, the mantle rock is rigid.

metamorphic rock Any rock mass that has been changed in composition or texture due to the action of heat, pressure, or chemically active fluids.

Moho The boundary between Earth's crust and mantle, where there is a sharp increase in the velocity of seismic waves, named in honor of Andrija Mohorovičić.

P–wave Primary seismic wave that is generated by earthquakes. It travels faster than an S–, or compression, wave and can pass through solids, liquids, and gases.

Pangaea The single supercontinent that existed about 250 million years ago when all the landmasses were merged into one. By about 200 million years ago, Pangaea had broken up and drifted apart as two separate land masses.

Panthalassa The planetwide ocean that existed at the time of Pangaea, meaning "all seas."

planetesimals Chunks of rock, metals, and ices that were formed in the early life of the Solar System some 4.6 billion years ago and that collected in ever-larger chunks that became the planets.

plate tectonics The movement of vast crustal plates accounting for much of Earth's geologic activity.

plume theory The idea that hot mushroom-shaped masses of magma rise out of the deep mantle and melt their way up through the crust as hot spots that produce volcanoes.

radioactive heating Heating within Earth, mainly within the crust, due to energy being released by radioactive elements such as uranium.

residual mountains Mountains that result from large, flat areas of land rising above the surrounding plain rather than folding up.

S–wave Secondary seismic waves called shake, or shear waves, that are generated by earthquakes. Unlike P– waves, S–waves travel slower and only through solids.

sedimentary rock Rock formed from clay, lime, sand, gravel, plant, or animal remains that have been squeezed together under great pressure or naturally cemented for long periods of time.

sediments Loose materials such as clay, mud, sand, gravel, lime, and other Earth materials eroded from rocks and deposited elsewhere by wind, water, and ice.

seismic waves Any of the classes of pressure waves generated by an earthquake, such as P–waves, S–waves, and surface waves.

seismologist Any scientist who studies seismic waves, such as pressure waves generated by an earthquake.

sill A long, wide, but thin body of intrusive igneous rock that lies parallel to the layering of the surrounding rock formations.

structure The way a substance is organized; e.g., a planet such as Earth being organized into an outer crust, a middle mantle layer, and a central core.

supernova The explosion of a giant star, during which all the elements heavier than iron are made.

surface waves Seismic waves that travel along Earth's surface and cause damage during an earthquake.

tsunami A gigantic, destructive ocean wave triggered by an earthquake in the seafloor. Tsunamis may reach heights of 100 feet (30 meters) or more.

turbidity current A mudslide down a continental slope. Mud may cascade down a slope at speeds up to 50 miles (80 kilometers) an hour.

volcanic mountains Mountains formed when molten rock flows up through vents in Earth's crust and solidifies as lava.

FURTHER READING

The following books are suitable for young readers who want to learn more about Earth's structure.

Cox, Reg. *The Seven Wonders of the Natural World*. New York: Chelsea House, 2000.

Nicolson, Cynthia Pratt. *Earth Dance: How Volcanoes, Earthquakes, Tidal Waves and Geysers Shake our Restless Planet*. Tonawanda, NY: Kids Can Press, 1999.

Watson, Nancy, et al. *Our Violent Earth*. Washington, DC: National Geographic, 1982.

WEBSITES

The following Internet sites offer information about and pictures of Earth, many of them with links to other sites.

http://www.extremescience.com/earthsciport.htm This site discusses, and shows photographs of, the most extreme instances of natural events, including the biggest earthquake, the highest volcanic eruption, and the deepest place in the ocean.

http://vulcan.wr.usgs.gov/Volcanoes/framework.html The site of the United States Geology Survey volcano observatory. Lists all of the known volcanoes in the world, with facts and photographs.

http://www.usgs.gov The site of the United States Geology Survey gives information on every aspect of Earth's structure, including earthquakes, floods, and maps. Includes fact sheets as well as maps and photographs.

BIBLIOGRAPHY

Bassett, William A. "What Is in the Earth's Core Besides Iron?" *Science*, vol. 266, December 9, 1994.

Bloom, Arthur L. *The Surface of the Earth*. Englewood Cliffs, NJ: Prentice-Hall, 1969.

Clark, Jr., Sydney P. *The Structure of the Earth*. Englewood Cliffs, NJ: Prentice-Hall, 1971.

Clayton, Keith. *The Crust of the Earth*. Garden City, NY: Natural History Press, 1967.

Dvorak, John J., Carl Johnson, and Robert I. Tilling. "Dynamics of Kilauea Volcano." *Scientific American*, August 1992, pp. 46–53.

Ernst, W. G. *Earth Materials*. Englewood Cliffs, NJ: Prentice-Hall, 1969.

Forte, Alessandro M., and Jerry X. Mitrovica. "Deep-mantle High-viscosity Flow and Thermochemical Structure Inferred from Seismic and Geodynamic Data." *Nature*, vol. 410, April 26, 2001, pp. 1049–1055.

Gallant, Roy A., and Christopher J. Schuberth. *Earth: The Making of a Planet*. Tarrytown, NY: Marshall Cavendish Corp., 1998.

Gallant, Roy A. *Exploring Under the Earth*. Garden City, NJ: Doubleday, 1960.

Guyot, François. "Earth's Innermost Secrets." *Nature*, vol. 369, June 2, 1994, pp. 360–361.

Leet, L. Don. *Causes of Catastrophe*. New York: McGraw-Hill, 1948.

Leet, L. Don, Sheldon Judson, and Marvin E. Kauffman. *Physical Geology*. Englewood Cliffs, NJ: Prentice-Hall, 1978.

Niu, Fenglin, and Lianxing Wen. "Hemispherical Variations in Seismic Velocity at the Top of the Earth's Inner Core." *Nature*, vol. 410, April 26, 2000, pp. 1081–1084.

Pinter, Nicholas, and Mark T. Brandon. "How Erosion Builds Mountains." *Scientific American*, special issue, 2000, pp. 24–29.

Skinner, Brian J. *Earth Resources*. Englewood Cliffs, NJ: Prentice-Hall, 1969.

Spencer, Edgar Winston. *The Dynamics of the Earth*. New York: Crowell, 1972.

Thatcher, Wayne. "Pop-Up Disaster." *Nature*, vol. 410, April 12, 2001, pp. 757–758.

INDEX

Page numbers in **boldface** are illustrations.

ABOUT THE AUTHOR

Roy A. Gallant, called "one of the deans of American science writers for children" by *School Library Journal*, is the author of almost one hundred books on scientific subjects, including the best-selling National Geographic Society's *Atlas of Our Universe*. Among his many other books are *When the Sun Dies*; *Earth: The Making of a Planet*; *Before the Sun Dies*; *Earth's Vanishing Forests*; *The Day the Sky Split Apart*, which won the 1997 John Burroughs award for nature writing; and *Meteorite Hunter*, a collection about his expeditions to Siberia to document major meteorite impact crater events. His most recent award is a lifetime achievement award presented to him by the Maine Library Association.

From 1979 to 2000, (professor emeritus) Gallant was director of the Southworth Planetarium at the University of Southern Maine. He has taught astronomy there and at the Maine College of Art. For several years he was on the staff of New York's American Museum of Natural History and a member of the faculty of the museum's Hayden Planetarium. His specialty is documenting on film and in writing the history of major Siberian meterorite impact sites. To date, he has organized eight expeditions to Russia and is planning his ninth, which will take him into the Altai Mountains near Mongolia. He has written articles about his expeditions for *Sky & Telescope* magazine and for the journal *Meteorite*. Professor Gallant is a fellow of the Royal Astronomical Society of London and a member of the New York Academy of Sciences. He lives in Rangeley, Maine.